Extreme Environmental Threats™

ANTARCTIC MELTING

The Disappearing Antarctic Ice Cap

Michael A. Sommers

The Rosen Publishing Group, Inc., New York

Pour les gens que j'aime au Canada . . .

Published in 2007 by The Rosen Publishing Group, Inc.
29 East 21st Street, New York, NY 10010

Copyright © 2007 by The Rosen Publishing Group, Inc.

First Edition

All rights reserved. No part of this book may be reproduced in any form without permission in writing from the publisher, except by a reviewer.

Library of Congress Cataloging-in-Publication Data

Sommers, Michael A., 1966–
Antarctic melting: the disappearing Antarctic ice cap/by Michael A. Sommers.—1st ed.
    p. cm.—(Extreme environmental threats)
Includes bibliographical references and index.
ISBN 1-4042-0741-4 (lib. bdg.)
1. Ice caps—Antarctica. 2. Global warming. 3. Sea level. I. Title. II. Series.
GB2597.S66 2007
363.738'74—dc22

                                                                2006000166

Manufactured in the United States of America

On the cover: Of the seventeen species of penguins in the world, only four live on Antarctica. They spend up to 75 percent of their lives in the icy Antarctic waters. **Title page:** Antarctic ice floes, also known as pack ice, form in gigantic and dangerous masses during winter and almost completely disappear in summer.

Contents

	INTRODUCTION	4
1	TERRA AUSTRALIS INCOGNITA	8
2	LIFE ON "THE ICE"	22
3	WARMING UP	35
4	MELTING DOWN	46
	GLOSSARY	56
	FOR MORE INFORMATION	58
	FOR FURTHER READING	60
	BIBLIOGRAPHY	61
	INDEX	62

INTRODUCTION

A giant petrel perches on a rock at Humble Island, near the Antarctic Peninsula. Petrels fish for krill and squid and feed on dead seals and penguins.

In early February 2005, scientists from around the world gathered in Exeter, Great Britain, for a climate change conference. Professor Chris Rapley, director of the British Antarctic Survey (BAS), a research organization that studies the planet's coldest continent, had a surprising announcement to make. Scientists monitoring Antarctica had discovered new evidence concerning the reduction of the continent's ice. Increased melting in the Antarctic Peninsula and West Antarctica was contributing to a rise in global sea levels to a greater degree than had been expected.

   Only four years earlier, in 2001, the Intergovernmental Panel on Climate Change (IPCC) had estimated that by 2100, world sea levels would rise between 4.3 and 30.3 inches (11–77 centimeters). Even the maximum increase of 30 inches might seem like a small amount, but such a change could have a major effect on coastal cities around the world, causing flooding, death, and destruction. At the time, scientists had predicted that Antarctica's contribution to rising sea levels would be small (about 5 percent). However, since then, studies have discovered that melting Antarctic ice caps are in fact contributing at least 15 percent to the current world sea level rise of 0.08 inches (2 millimeters) a year. What was once known in scientific circles as the world's "sleeping giant" was now, according to Rapley, beginning to wake up.

The world's coldest, highest, driest, windiest, and most mysterious continent, Antarctica contains 70 percent of the world's freshwater and 90 percent of its ice. Antarctica is about the size of the continental United States and half of Mexico combined, and it is the coldest place on earth, with average temperatures of −70 degrees Fahrenheit (−57 degrees Celsius). Not even the active volcanoes that erupt from time to time have much warming effect on this continent, 98 percent of which is ice. Frigid air temperatures mean that snow rarely falls. Even in summer, temperatures are far below freezing in most of Antarctica. They seem colder still due to the fierce katabatic winds (*kata* is Greek for "downward") that sweep down from the high central regions to the coast at 125 miles (200 kilometers) per hour. While much of its frozen landscape is snow covered, precipitation is so scarce that Antarctica's interior is actually the world's biggest desert. No flowering plants are able to grow, and the largest land animals are microscopic.

Unlike the Arctic, where ice floats on top of the ocean, Antarctica's ice, which is 3 miles (4.8 km) thick in some places, sits upon a landmass, most of which is extremely elevated. The South Pole itself is on a plateau

INTRODUCTION

10,000 feet (3,050 meters) above sea level. Antarctica's highest peak, the Vinson Massif, is over 3 miles (4.8 km) high. If, one day, this icy continent did begin to thaw, the consequences would be catastrophic. If all of Antarctica's ice melted, sea levels around the globe would rise about 200 feet (60 m)! This would flood coastal cities all over the world and cover many island chains and large continental areas with water. Fortunately, for the time being, it is virtually impossible for this to happen in a land where the temperature almost never reaches above freezing. No wonder scientists who carry out research there refer to it as "the Ice."

Nevertheless, something alarming is occurring in some parts of Antarctica. As a result of global warming, the ice shelves that surround the continent's edges are becoming weaker, cracking, and breaking off into the ocean. As soon as this ice falls into the water, the sea rises a little bit. How much the world's oceans might rise due to Antarctic conditions is a hotly discussed topic in government and scientific circles and in the press. There are various possible scenarios and no easy answers. One certainty is that the discussion is focusing attention on the world's least known but crucially important continent. It is a place where ice holds answers to questions about the earth's past, present, and future.

1 TERRA AUSTRALIS INCOGNITA

A full moon and long exposure time allowed this photo to be taken during the six-month-long Antarctic night.

Antarctica wasn't always covered with ice and snow. In fact, millions of years ago, it possessed an almost tropical climate. Originally, Antarctica, Australia, Africa, and South America, along with India and the island of Madagascar, were all part of a supercontinent in the Southern Hemisphere known as Gondwana. If each of these regions were a separate jigsaw-puzzle piece, their edges would fit together neatly. Other evidence proves that these landmasses were formerly joined together. Over the years, geologists have found identical

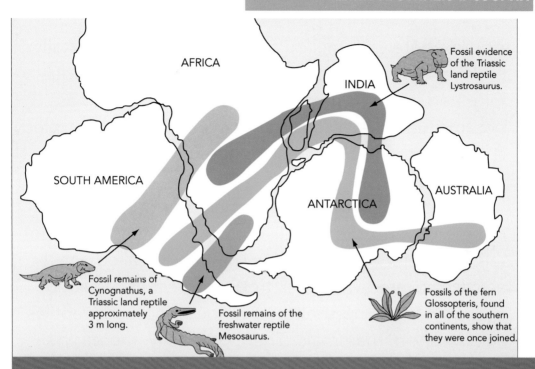

This illustration shows how India and the four continents of today's Southern Hemisphere originally fit together as the earth's megacontinent, Gondwana. There are various theories about the geographic location of the supercontinent. Most geologists agree that it was situated close to the South Pole for about 200 million years before moving farther north.

rocks as well as plant and animal fossils of the same age on all the southern continents. There are also mountain ranges that are shared by more than one continent. For example, the Ellsworth mountain range in western Antarctica continues into South Africa and South America.

About 160 million years ago, Gondwana began to split up. Africa and Madagascar were the first to break away, followed by India, New Zealand, and then Australia, which separated around 60 million years ago.

Approximately 25 million years ago, Antarctica finally broke away from South America. Left alone, it gradually migrated south to its present position. Its landmass became further cut off as strong ocean currents encircled it, leading to increased isolation from other southern continents.

Until the middle of the Tertiary period (an interval of time lasting from 65 million to 1.8 million years ago), the "white continent," as it is sometimes called, was actually covered with thick tropical forest. It provided abundant food and shelter to various dinosaurs, reptiles, and, later on, mammals. Then, gradually, the earth's climate became colder. Plunging temperatures resulted in an ice age that caused glaciers to form, particularly at the poles.

ANTARKTOS

The first reference to the Antarctic appeared in ancient Greece sometime around 400 BC. Greek philosophers believed the earth to be an immense sphere, which was necessarily stabilized by two large landmasses at its north and south poles. The Greeks referred to the north as Arktos, which is Greek for "bear," because they located the north by searching the sky for the Great Bear constellation of stars. It is believed that the philosopher Aristotle first coined the term "Antarktos" to refer to the

TERRA AUSTRALIS INCOGNITA

vast territories that the Greeks imagined lay clustered around the South Pole.

For the next 2,000 years, humans attempted to prove that Antarctica actually existed. In the second century AD, the Greek geographer and astronomer Ptolemy (85–165) referred to Antarctica as Terra Australis Incognita (Unknown Southern Land). He believed it to be a fertile but uninhabited land that couldn't be reached without traveling through regions full of terrible dangers and frightening monsters. For centuries afterward, similar myths about Antarctica abounded. In the Middle Ages, however, the Catholic Church acquired great influence in Europe. According to the church, God would never have created a land that was so terrible and desolate. Adhering to the church's views that such a place couldn't be real, over the next 300 years, people gradually forgot about the existence of Antarctica.

The fifteenth and sixteenth centuries marked an era of great explorations and discoveries. Navigators such as Vasco da Gama (1460–1524), Ferdinand Magellan (1480–1521), and Amerigo Vespucci (1454–1512) sailed around the world and revolutionized the study and knowledge of the earth's geography. When Magellan rounded the tip of South America in 1520, he passed through an ocean strait (which today bears his name), beyond which snow and ice appeared. As a result, on their world maps, geographers placed the legendary

Antarctic Melting: The Disappearing Antarctic Ice Cap

"white continent" at the Pacific Ocean's southern extreme.

In the late 1600s, southern expeditions increased, as did sailors' tales of being caught in terrible storms off Cape Horn (the tip of South America). The storms blew their ships toward bitterly cold, harsh lands covered with mountains of snow and ice. Sailors' dramatic accounts fueled other expeditions as European explorers dreamed of staking claim to a mysterious continent. Nonetheless, it wasn't until 1772 that Captain James Cook (1728–1779), a British explorer and navigator, sailed directly south from Cape Town, South Africa, in search of territory for England. After getting stuck in ice and dodging icebergs, he succeeded in crossing the Antarctic Polar Circle (at latitude 66°33′) in 1773. In doing so, he was the first to finally set eyes on and describe the shores of the Antarctic continent.

Captain James Cook was known for his expert navigating skills. Despite passing through heavy storms and dangerous seas, his ship survived the three-year Antarctic expedition without the loss of a single man.

This engraving from the State Library of Tasmania, New Zealand, shows an Antarctic seal hunt in 1838. In the early 1800s, "seal wars" were waged as European merchants competed to acquire the greatest number of the animals' valuable skins. Over three million seals were killed each year during this time period.

Cook's logbook descriptions had an enormous impact. They destroyed centuries of myths about a fertile Antarctic continent filled with abundant natural riches. Instead, it appeared that Antarctica was frigidly cold, inhospitable, and lifeless. Of great interest, however, were Cook's reports of numerous seals and whales living in the freezing southern ocean waters. Over the next forty years, thousands of ships set sail for Antarctic waters, not to explore new lands, but to hunt whales and seals, whose precious oils and glossy pelts were highly prized in Europe. Antarctica itself became less feared and better known as crews set up temporary bases on ice-covered islands long enough to slaughter these animals at an alarming rate. In search of greater profits, hunters traveled farther south. Among them were the first men to actually disembark upon the southern continent. John Davis, an American seal hunter, is largely believed to have been the first person to ever set foot on Antarctica, in 1821. However, he was only interested in killing seals, not in discovering new territory for his country.

The race between nations wanting to lay claim to Antarctica began shortly after John Davis's landing. In 1840, two explorative expeditions—one French, the other American—planted their nations' flags upon the Antarctic coast. The quarrel as to who planted their flag first was never resolved, but it increased other nations' desires to conquer Antarctica. A few months later, British navigator Sir James Clark Ross (1800–1862) braved ice blocks and icebergs and found his way to an inland sea (to which his last name was subsequently attached). After planting the British flag on Antarctica, Ross discovered the continent's 12,447-foot (3,794 m) volcano, Mount Erebus. (Still active, its crater is a permanent lake of lava.) His exploration came to an end, however, when he found his path blocked by an immense cliff of sheer ice, known today as the Ross Barrier.

In spite of Ross's discoveries, the scarcity of seals and whales due to decades of slaughter, coupled with the impossibility of colonizing such a cold region, led to diminished interest in Antarctica during the rest of the nineteenth century. Yet, even if Antarctica possessed little territorial or economic value, its unstudied secrets intrigued scientists.

IN THE NAME OF SCIENCE

Adrien de Gerlache (1866–1934) was a young Belgian navy officer who liked the cold and wanted to advance science by studying the mysteries of the globe's last unexplored land. Determined to organize an expedition to Antarctica, he held fund-raising parties and concerts across Belgium in order to raise money for his journey. In 1897, on a small boat laden with scientific equipment, de Gerlache finally set sail for the Southern Hemisphere. He was accompanied by a group of young scientists (their average age was twenty-eight) that included a zoologist, botanist, geologist, meteorologist, oceanographer, and photographer.

Arriving at the Antarctic Peninsula, de Gerlache and the other scientists explored coast after coast. They took notes and measurements; drew maps; and collected samples of rocks, ice and seawater, and plant and animal life. Heading farther and farther south, the expedition

How Time Flies in Antarctica

The South Pole has six months of endless light and six months of nonstop darkness. On September 23, the sun rises at the South Pole, marking the beginning of six months of daylight. At the end of March, the sun goes down and is not seen again for another six months. At the South Pole, the one sunrise and one sunset of the year can each last for several days.

Telling time is tricky in a place where all time zones come together. In order to give a simple answer to the question "What time is it?" Antarctica officially follows New Zealand time.

Members of the U.S. Antarctic Program help dock a ship at Palmer Station, on Antarctica's Anvers Island. With a large laboratory and a seawater aquarium, the station allows scientists to study birds, seals, and other marine life. In the winter, the station hosts about ten scientists.

was eventually trapped by ice. Terrified, the crew found themselves forced to spend the entire winter in Antarctica. As the sun disappeared and the unending polar night set in, survival became difficult for the men. The air was often −4°F (−20°C). One crew member died and many became sick. Georges Lecointe, the expedition's second in command, wrote in his book *In Penguin Country*, "Our features are drawn, our faces are lined, our skin is green and our eyes dull and lifeless. It only needed 1,600 hours of uninterrupted night to turn us all into old men." In order to stay alive, the men had to eat raw penguin and seal, whose oily flesh disgusted them. Finally, in mid-March of 1899, nearly thirteen months after they were first trapped, the ice melted and separated enough to allow the expedition free passage to the open sea.

Scientists were thrilled with the observations and newly discovered specimens that de Gerlache's team brought back to Europe. In spite of their misadventures, the first scientific voyage to Antarctica was ultimately a great success. The knowledge that it was possible to withstand the brutal Antarctic winter would inspire many others throughout the twentieth century. Two famous expeditions by competing explorers, Norway's Roald Amundsen (1872–1928) and Britain's Robert Falcon Scott (1868–1912), focused on the search for Antarctica's heart: the South Pole.

The Great Race to the South Pole

The contest to discover the South Pole is one of history's great races. The winter of 1911 found Amundsen's and Scott's expedition teams within 62 miles (100 km) of each other on the Ross Ice Shelf. From their base camps, they prepared for the journeys that would take them across 9,320 miles (15,000 km) of unknown territory with only compasses to guide them. Their mission was to reach and claim discovery of the South Pole. Amundsen's team made their equipment as light as possible: from tents and skis to lightweight sleds pulled by dogs. Meanwhile, Scott and his team's tactic was to rely on powerful but heavy gear, such as motorized sleds.

Scott's team set off on November 1, 1911. Almost immediately, the expedition ran into trouble. Weighing 660 pounds (300 kilograms) each, the team's sleds, pulled by ponies, kept getting stuck in the snow. So did the ponies, and they eventually had to be slaughtered. In addition, the motorized sleds kept breaking down. The team's worst nightmare, however, was realized when they finally reached the South Pole. They found the Norwegian flag planted there, along with a letter from Amundsen. Traveling lighter and faster (with the aid of Greenland huskies), Amundsen's four-man team had reached the South Pole forty-four days earlier (on December 14) after a smooth, uneventful journey. The British team's return to home base proved harrowing as the men battled severe blizzards, frostbite, scurvy, and gangrene that caused their limbs to grow black. Only 11 miles (18 km) from their final destination, Scott and his two remaining teammates ran out of fuel and food. Eight months later, a patrol team discovered all three men frozen to death in their tent. They also found a diary entry written by Scott that began: "Great God! This is an awful place!"

TERRA AUSTRALIS INCOGNITA

Upon arriving at the South Pole, Roald Amundsen's team spent three days calculating to confirm they had arrived at the earth's southernmost tip. They planted the Norwegian flag in the ice *(above)* as proof of their historic undertaking. Amundsen traveled on skis *(top left)* that were made as light as possible by the expedition's carpenter. The skis allowed Amundsen and his team to cover distances of up to 31 miles (50 km) a day. A chocolate bar *(bottom left)* was recovered from Robert Falcon Scott's ship *Discovery* after its return from Antarctica. The chocolate was preserved in a cigarette tin.

An explorer in the early 1900s tries on clothing that a London company manufactured specifically for use on Antarctic expeditions.

Amundsen's and Scott's exploits made front-page headlines around the world and launched a permanent interest in the Antarctic. Nations including the United States, Germany, Australia, and New Zealand all launched scientific and discovery expeditions. Airplanes, balloons, and snowmobiles crossed the continent. People scaled its mountains and bored into its depths with powerful drills. They charted, mapped, and photographed its features from land and by air. Weather stations, temporary laboratories, and permanent scientific camps sprung up. By 1958, more than sixty bases and stations were operating in Antarctica. The largest base, McMurdo, was built by the United States. Since its construction, it has hosted American scientists interested in studying Antarctic erosion, deep ice, biology, geology, meteorology, oceanography, and human behavior in hostile environments.

McMurdo Station, the American research base in Antarctica, is surrounded by volcanic mountains. With about 1,000 scientists living and working there, McMurdo is Antarctica's largest "town," with over 100 structures, including a harbor, airport, and helicopter pad. Buildings are linked by aboveground water, sewer, telephone, and power lines.

Scientific research in the Antarctic has become more technologically advanced in recent years. With the scientific community increasingly certain that Antarctica has a major influence on the earth's climate and ecosystems, individual national studies have given way to major projects involving researchers from many nations working together. One of their major concerns is the melting of Antarctic ice.

2 LIFE ON "THE ICE"

Along the endless white Antarctic coast, it is sometimes difficult to tell where the land ends and the sea begins.

In Antarctica today, scientific groups from twenty nations operate thirty-eight permanent, year-round stations. Another eight nations run thirty-four summer-only stations. As a result, Antarctica's population fluctuates from around 1,000 people in the winter to over 4,000 during the summer months.

These stations function like small, self-contained villages with satellite links to the outside world. Over the years, technological advances have improved polar travel and living. Lightweight snow vehicles

A vast dome acts as a protective shell for the buildings at the Amundsen-Scott Station, an American research base located in inland Antarctica. The station was originally built at the South Pole in 1956, but over time it has moved with the constantly shifting ice sheets upon which it was constructed.

have replaced husky-drawn sleds, aircraft outfitted with skis take scientists to and from their fieldwork, and strengthened ships with onboard laboratories navigate through thick ice. As well as having all of the latest scientific tools and gadgets, many of these stations offer residents a surprising number of modern comforts, including highly equipped kitchens, libraries, saunas, sport facilities, and even cinemas and bowling alleys. Meanwhile, much of the information the public receives about Antarctica comes from several hundred miles up in space. Various satellites track changes in water temperature; sea ice; and the ozone layer, a concentration of the ozone (O_3) molecule in the earth's atmosphere. The ozone layer protects the planet from the harmful effects of the sun's ultraviolet (UV) rays. Increasingly, however, this natural shield is being damaged by human-produced pollution.

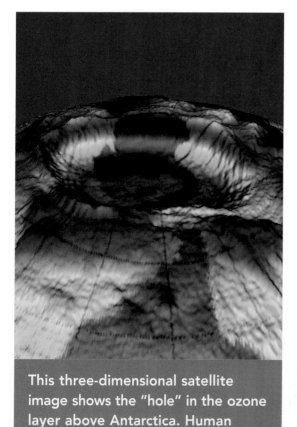

This three-dimensional satellite image shows the "hole" in the ozone layer above Antarctica. Human activity has been largely responsible for destroying the ozone layer.

A SCIENTIFIC ZONE

Even though a number of adventurers have planted flags on various parts of Antarctica in the name of their homelands, the continent doesn't belong to any nation. To travel there, no passport or visa is required. Britain, Norway, Chile, Argentina, France, New Zealand, and Australia have all historically staked possession of some or all of Antarctica. Yet, so far their claims have been rejected. In 1959, the Antarctic Treaty was signed by twelve member nations. Along with the seven claimant nations listed above, these member nations included the United States, South Africa, Japan, Russia, and Belgium. The treaty stated that the continent would be administered by all consultative (decision-making) members. As of 2004, there are forty-five treaty member nations: twenty-eight are consultative, a status they received because they "conduct substantial research [on the continent]," and seventeen are non-consultative.

The Antarctic Treaty basically officialized Antarctica as a scientific zone where specialists from all nations could freely study the white continent in compliance with certain regulations. These rules included no military activity, nuclear experimentation, or economic exploitation of natural resources. Over time, further agreements were signed relating to the preservation of wildlife and natural resources and the management of sea and land pollution. For example, nations are responsible for removing all waste generated at their stations and labs in order to keep Antarctica's environment pure. (Unfortunately, although this is a rule, it is not always followed.) In 2004, a permanent Antarctic Treaty Secretariat was founded in Argentina's capital, Buenos Aires. Its role is to document and report on the annual meetings of Antarctic Treaty members and keep historical

Antarctic Melting: The Disappearing Antarctic Ice Cap

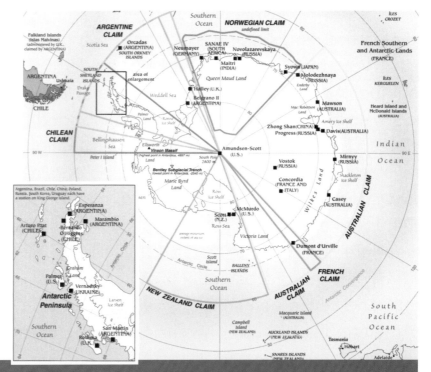

This map shows a detailed view of the Antarctic continent. Although many of Antarctica's consultative member nations have made claims to Antarctic territory, none of these claims have been recognized. The small black squares represent year-round scientific research stations.

records. Consultative members continue to meet every year to make decisions about how the territory should be managed.

THE LAY OF THE LAND

The Antarctic ice sheet is the largest single mass of ice on the planet, covering an area close to 5.4 million square miles (14 million square km). Split by 2,458 miles

(3,956 km) of the Transantarctic Mountains, it is generally divided into three parts: East Antarctica, West Antarctica, and the Antarctic Peninsula.

East Antarctica and West Antarctica are considered two separate polar ice sheets. East Antarctica is the largest and least explored part of the Antarctic. Aside from a few scientific expeditions and exploratory ventures, hardly any humans have set foot on its cold and hostile territory. The coldest temperature known in the world, –129.2°F (–89.6°C), was recorded there in 1983, at the Russian station of Vostok. East Antarctica's ice sheet contains about 85 percent of the continent's ice and sits at a high elevation on top of a major landmass. Numerous regions, including the South Pole, are roughly 10,000 feet (3,000 m) above sea level.

In contrast, the majority of West Antarctica's ice sheet, which is 11 percent of the continent's ice, lies as deep as 8,200 feet (2,500 m) below sea level. Because of West Antarctica's flatter, more accessible terrain, most nations have set up their weather stations and scientific bases there, especially along the coast.

Representing only 4 percent of the continent's territory, the Antarctic Peninsula resembles an arm reaching toward South America. It is the land closest to Chile and Argentina. As such, it receives a growing number of adventurers and tourists who are fascinated by the untouched Antarctic wilderness. It is the warmest region of Antarctica and the only area where, for at least two months out of the year, temperatures rise above freezing.

Types of Ice

Since polar temperatures are almost always freezing, the snow that does fall upon Antarctica's ice sheets hardly ever melts. Instead, year after year, snow crystals accumulate, pressing down on the crystals beneath them. Over time, this pressure transforms them into very dense glacial ice, also known as a glacier. Having accumulated over some 30 million years, it is hardly surprising that Antarctic ice—the oldest ice in the world, beating out Greenland, the Himalayas, and the Alps—weighs billions of tons.

    Glacial ice appears to be frozen solid. However, when placed under tremendous pressure, lower layers of ice will begin to flow like a thick, oozing liquid. To help visualize what this is like, imagine liquid ice as toothpaste that is being squeezed out of a tube as a result of pressure. The force of gravity causes underground liquid ice to slowly move out toward the sea. The slippery ice sheet itself acts like a conveyor belt that transports the ice to the ocean. While the surface of the ice sheet, exposed to icy winds and freezing temperatures, remains cold, the bottom of the sheet (closer to the center of the earth's hot core) is often

An aerial view of Antarctica's Taylor Glacier, located on the East Antarctic ice sheet. Taylor Glacier is an arching stream of ice about 35 miles (54 km) in length that flows in a U shape around rocks as it winds its way through a low-lying region known as the Dry Valleys.

warmer. At 3,200 feet (roughly 1,000 m) below the surface, liquid ice and even water can form. Creating underwater channels, the water flows toward the sea, carrying chunks of ice with it. These quick-moving channels are known as ice streams. Some ice streams are up to 30 miles (48 km) wide and hundreds of miles in length.

On the edges of ice sheets, on or near Antarctica's coasts, ice (which is less dense than seawater) begins to float. While still attached to the continental ice sheet, this ice is known as an ice shelf. Much of Antarctica's coastline consists of ice shelves, which are enormous slabs of floating ice that poured out of glacial ice sheets and onto the ocean. Some are the size of countries. With a surface area of 208,500,000 square miles (540,012,500 sq km), the famous Ross Ice Shelf is the size of Great Britain and the largest shelf in the Antarctic. It contains a third of all of the continent's floating ice. At the base of an ice shelf, ice and seawater come into contact. If the ocean is warm, ice from the shelf's bottom will melt, adding cold freshwater to the sea. This mass, which is known as Antarctic bottom water, is heavier than normal seawater and forms a cold current that flows along the bottom of the ocean at depths of more than 9,843 feet (3,000 m).

Occasionally, erosion and change in temperature create fractures and cracks in the ice shelves. Over time, these fractures cause enormous chunks of ice to break off from the shelf and fall into the ocean. These great floating blocks are called icebergs. Since the beginning of polar navigation, icebergs have been responsible for many catastrophes. There are tens of thousands of icebergs around Antarctica. Some weigh up to 400 million tons (360 million metric tons) and reach heights equivalent to that of a ten-story building. If melted, such an iceberg would furnish a year's worth

LIFE ON "THE ICE"

Towing Icebergs

It is estimated that all of Antarctica's icebergs, if melted, could supply a third of the world's annual water supply. With so many nations confronting droughts and water shortages, the possibility of great amounts of free freshwater is appealing. The only problem is transportation.

Over the past century, there have been some imaginative attempts to overcome this obstacle. In the 1950s, American oceanographer John Isaacs of the Scripps Institution in Jolla, California, calculated that six powerful tugboats could haul an 18.5-mile-long (29.8 km) iceberg from the Antarctic to California in only a few months. In the 1970s, two American scientists, John Hult and Neil Ostrander, worked out a scheme for joining various icebergs together, carving their fronts to resemble the bows of ships (in order to move with greater ease through the water), and towing the entire iceberg fleet with nuclear-powered boats. Perhaps the wildest iceberg-towing scheme, however, was that of Saudi Arabian prince Mohammed al Faisal. In 1977, he created a company whose goal was to bring a 100-million-ton (90.7 million metric tons) iceberg 8,700 miles (14,000 km) to the Saudi Arabian port of Jedi. Prior to one of the many public conferences he gave on the subject, he had a helicopter "rescue" a 2-ton (1.79 metric tons) iceberg from an Alaskan glacier. Broken into cubes, the 10,000-year-old ice was then used to refresh the drinks of 200 participants at the conference, held on the Ames University campus in Iowa.

Unfortunately, numerous considerations, such as the time required to tow the heavy ice, the amount the iceberg would melt, and problems of how to deal with the iceberg at its destination have, so far, kept these projects in the planning phase.

of freshwater for a city of 3 million residents! One of the biggest icebergs ever seen was sighted in 1956 by the crew of an American ship called USS *Glacier*. With a length of 208 miles (335 km) and a width of 60 miles (97 km), it was around the size of Belgium.

THE SOUTHERN OCEAN

Surrounding the Antarctic continent is the Southern Ocean. Formerly called the Antarctic Ice Ocean, it connects the Atlantic, Pacific, and Indian oceans. Representing 20 percent of the surface area of the earth's oceans, it is the only ocean that circulates around a continent without any obstacles. Although extremely cold and covered with ice for most of the year, it also supports a surprising variety of marine life. This includes microscopic algae and plankton, fish, squid, and krill, which are tiny shrimplike creatures that are the favorite food of Antarctic birds, seals, and whales.

Salty ice, or sea ice, forms on the Southern Ocean. It originates at the edge of the continent and is blown by strong winds into the ocean, forming ice packs (also known as ice floes). The sea ice is constantly moving and changing in shape and size, and as temperatures fall, it can reach an average thickness of 10 to 15 feet (3 to 4.6 m). As the ice expands and moves (driven by fierce winds, an ice floe can travel over 37 miles (60 km) in

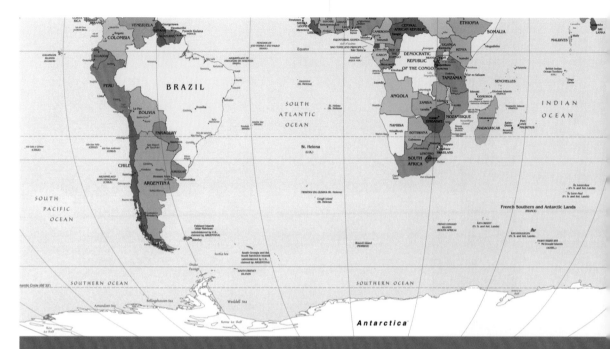

This map shows how the Southern Ocean links the Atlantic, Indian, and Pacific oceans. The average depth of the Southern Ocean is 2.5 miles (4 km). Its surface covers 7.8 million square miles (20.3 million sq km), which is about twice the size of the United States. It also contains the most powerful current in the world, the Antarctic circumpolar current.

one day), it can be a hazard for ships, trapping and even crushing them. Each year, the ice surrounding the continent grows from about 1.16 million square miles (3 million sq km) at the end of summer to 8 million square miles (roughly 20 million sq km) at the end of winter. By the time summer comes around again, close to 85 percent of the ice will have melted back into the sea.

The Southern Ocean, also known as the Antarctic Ocean, is considered by oceanographers to be the heart of the world's oceans. It functions like a powerful engine, pumping great quantities of cooler water into the Pacific, Atlantic, and Indian oceans, which oxygenates these oceans. The Southern Ocean also has an important effect on the Antarctic climate. With water temperatures that range from 28.8° to 50°F (–1.8° to 10°C), it supplies more warmth than the sun. This directly affects regional plant and animal life as well as the rate at which ice forms and melts. Scientists consider study of the Southern Ocean crucial to understanding future changes in the earth's climate.

3 WARMING UP

> The Weddell Sea ice shelf is seen here on a crystal-clear Antarctic day.

Aside from seeking answers about the Antarctic continent itself, scientists study Antarctica in order to understand the rest of the planet. For instance, Antarctic ice functions as a sort of global air conditioner. Besides cooling the earth's air, it also removes humidity from it since the precipitation that falls from the atmosphere to the ground becomes solid. If ice began to melt in increasing quantities, hotter summers could result throughout the world. In order to understand global changes that are taking place now and those that are likely to happen in the future, it is essential to track and analyze changes that are occurring in and around Antarctica.

GLOBAL WARMING

In recent years, many people have become increasingly concerned about a phenomenon scientists refer to as global warming. Studies have shown that average temperatures are consistently rising in many places. Data shows that this warming trend has been happening during the last 150 years. Many scientists believe that one factor responsible for global warming is the increase of greenhouse gases in the atmosphere. These gases function like a greenhouse by trapping heat on the earth and preventing it from escaping into space.

One of the major greenhouse gases is carbon dioxide (CO_2). All animals, including humans, produce carbon dioxide. After breathing in oxygen from the air, they exhale carbon dioxide. The carbon dioxide works like a sponge to absorb heat from the earth's surface. This natural greenhouse effect helps to make the earth warm enough for life to exist. Problems occur, however, when too much carbon dioxide is produced. This can cause the earth's climate to heat up.

Human activities are a major source of this overheating. Waste gases produced by the burning of fossil

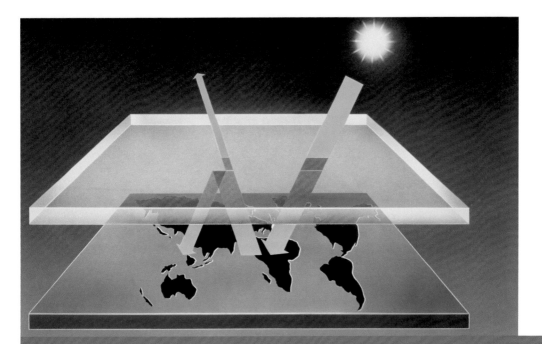

This diagram illustrates how the greenhouse effect operates. The purple layer represents the troposphere—the part of the earth's atmosphere where greenhouse gases are building up. Because of these gases, an increased amount of solar radiation is being trapped, allowing only a fraction of reradiated solar energy (the yellow arrows) to return to space. The rest of this energy rebounds back to the earth. As a result, over time the planet is gradually growing warmer.

fuels such as coal and petroleum are causing carbon dioxide and other greenhouse gases to be released into the atmosphere like never before. Global warming can affect many features of the climate: temperatures, wind patterns, precipitation, and severe weather events such as tidal waves and hurricanes. Some results of global warming that have already been observed include

Global warming may have played a part in one of the greatest natural catastrophes in American history. Hurricane Katrina swept through Gulfport, Mississippi, on August 29, 2005 (above). The hurricane left more than 1,400 people dead and around $75 billion in estimated damage. It also destroyed large areas of the historic city of New Orleans, Louisiana.

retreating Alpine glaciers, rising sea levels, and increased melting of the polar ice caps.

To measure changes in levels of greenhouse gases, it is necessary to be as far away as possible from all plants, animals, and human activity. Unsurprisingly, the best site in the world is Antarctica. Since 1956, scientists have been tracking global levels of carbon dioxide and other greenhouse gases from a station at the South Pole.

WARMING UP

Hot Statistics

- The 1980s and 1990s were the warmest decades on record.
- The ten warmest years since global weather records first began occurred in the past fifteen years.
- The twentieth century was the warmest century in the last 600 years.
- Over the last 100 years, the earth's temperature has risen by 0.9°F (0.5°C) and global sea levels have risen by 6 to 8 inches (15 to 20 cm).
- Most scientists believe that average global temperatures could rise by 4.8° to 6.3°F (1° to 3.5°C) over the next century.

Global warming has had almost no effect on the Antarctic continent because of the frigid temperatures of East and West Antarctica. To date, the average temperature on the Antarctic continent is −70°F (−57°C). However, significant changes have begun to take place on the Antarctic Peninsula, particularly in the last ten years.

Higher temperatures are responsible for increased breaking off and melting of some icebergs. However, if an iceberg melts, this does not increase sea levels. A floating iceberg displaces seawater equal to its own weight and volume, so when it melts, the sea level is not affected. This rule—known as Archimedes' principle—is what allows submarines to stay afloat. Only ice that breaks off or melts from land can cause sea levels to

Archimedes' Principle

Archimedes (287–212 BC), a Greek philosopher and astronomer, is considered one of history's greatest mathematicians. One day, when he was taking a bath, Archimedes made the following observation: the deeper he submerged his body in the water, the higher the water level rose, and the lighter his body felt. His conclusion—that an object partially or totally immersed in a fluid is buoyed up by a force equal to the weight of the displaced fluid—became known as Archimedes' principle. It explains not only the buoyancy of icebergs, ships, and other vessels in water, but also the apparent loss of weight of objects when underwater.

rise. Indeed, increased warmth may be weakening ice shelves, causing cracks that lead to ice breaking off into the water. As soon as the ice that was attached to the continent falls into the ocean, the sea rises a little. It is difficult to predict how much sea levels could increase as the result of global warming. Some scientists estimate that by 2100 ocean levels around the world might rise by as much as 3.3 feet (1 m).

A rise in sea levels could actually lead to further breaking up and melting of the Antarctic ice sheets. Although there is much scientists still don't know about West Antarctica, they believe that global warming and rising seas could cause warmer water to flow beneath the low surface of its ice sheets (most of which are below sea level). This could cause enormous chunks of ice—the size of the states of California or Texas—to break off into the water. Such an occurrence could cause seas around the world to rise as much as 18 feet (5.5 m), flooding coastal cities such as Boston, Barcelona, and Rio de Janeiro. Many of these cities would have to be abandoned. Coastal farmland would be destroyed and people would be without homes, food, and livelihoods. However, most scientists consider such an event highly unlikely. Even if it did happen, it would take thousands of years. Meanwhile, some glaciologists argue that global warming might cause more moisture to evaporate and then return to earth in the form of increased precipitation. In Antarctica, this would lead to more snowfall, which in turn would cause ice shelves to grow and even expand.

GLACIOLOGY

Glaciologists are scientists who specialize in the study of ice. In Antarctica, they are involved in many scientific activities. They monitor when a giant iceberg breaks off from a continental shelf and track its progress away from the coast. On the shelves themselves, they track how quickly ice streams are moving and how much ice is being transported to the sea. They also measure the thinning of ice due to the melting that takes place on its underside. Their observations help them to predict how long it will

French glaciologist Bernard Francou checks a control mark used to calculate the amount of ice that is melting from the 17,500-foot (5,345 m) Chacaltaya Glacier in the Bolivian Andes. It is estimated that by 2015, its snowy caps will have completely disappeared. By studying melting trends on glaciers in Antarctica and the rest of the world, scientists can foresee the effects of global warming and predict future climate changes.

take for certain ice sheets to break off and collapse into the sea. Glaciologists are working to create computer models that will accurately predict continental ice loss. By entering a series of precise field measurements—the ice thickness, air temperature, and speed at which ice moves over time, for example—into a computer program, scientists can calculate probable future activity (such as the amount of glacial melting in a given place over a specific period of time).

Glaciologists rely on various sources besides field surveying for their information. Ice-penetrating radar can track the flow and measure the thickness of liquid ice and ice streams beneath the surface of an ice sheet. Space

Drilling for Ice

Siple Dome is a small hill in West Antarctica's ice sheet that rises about 1,000 feet (300 m) above the surrounding ice. It is named after Paul Siple, the Antarctic explorer who in the 1940s discovered the principle of the windchill index (the effect that the speed of cold wind has on a person's loss of body heat).

Over the course of three summers in the late 1990s, two teams of drillers worked twelve-hour shifts, seven days a week, in order to remove ice cores that went all the way down to Antarctica's rocky bottom, around 3,300 feet (1,006 m) beneath Siple Dome's icy surface.

Drilling was done from a 110-foot (33.5 m) tower on the ice. The complex equipment, which included a powerful motorized drill and a bucket to hold the cores, was 65 feet (19.8 m) long and lowered into the ice from a hole in the tower. Each core that was removed measured 3 feet (0.9 m). As the cores emerged from the ground, they made a popping sound—not unlike the opening of a champagne bottle—as bubbles of high-pressure air that were trapped in the ice exploded. Ice that couldn't be used inspired drillers, technicians, and glaciologists to create a new tradition: toasting the day's success with cocktails whose ice cubes were 10,000 years old.

Workers remove a piece of ice core drilled from beneath the Siple Dome in 1998. Drilling was completed in January 1999 when the final 3-foot (0.9 m) core was removed from the bottom of the ice sheet. By examining air molecules trapped in the ice, scientists are able to study the earth's past climate. The ice from the core probably fell as snow on Antarctica over 80,000 years ago, when the planet was gripped in a long series of ice ages.

satellites can take very accurate pictures of Antarctica. In doing so, they can also track the movements of ice sheets, shelves, and icebergs, establishing patterns that can help scientists predict melting and collapses.

Meanwhile, locked deep below Antarctica's surface lie records of the earth's past and clues to the changes it might undergo in the future. To get at this information,

WARMING UP

glaciologists use powerful drills that bore deep into ice sheets. The ice samples that emerge are called ice cores.

Ice cores are important because within their layers, they preserve records of precipitation, temperature, and the presence of substances in the atmosphere including gases; soot; industrial pollutants; ash; aerosol particles; and dust from deserts, volcanoes, and meteors. If a toxic cloud from an explosion in North America eventually made its way to Antarctica, particles would mix with falling snowflakes. When the fallen snow turned to ice crystals, the traces of the toxic cloud would be preserved inside the crystals and trapped air bubbles. For example, traces of the 1883 eruption of the Krakatau volcano on an island in Indonesia and radioactive fallout from nuclear testing in the United States during the 1950s have turned up in Antarctic ice samples. Examining these cores permits glaciologists to examine past conditions with great accuracy. A core taken from a depth of 1 foot (0.3 m) might reveal information about the climate 50 years ago, while drilling 1,312 feet (400 m) down into the ice offers a look at conditions 10,000 years ago.

4 MELTING DOWN

Gentoo penguins on the Arctowski Peninsula in Antarctica rest between dives for their meals of fish and seafood.

On February 25, 1995, glaciologists from the British Antarctic Survey announced that two major and quite shocking events had occurred along the Antarctic Peninsula. First, an immense iceberg had just been created, having abruptly broken away from the Larsen Ice Shelf. Measuring 927 square miles (2,401 sq km) and known as A25, it was roughly half the size of the state of Delaware. Secondly, an ice shelf that had always linked the Antarctic coast to Ross Island had collapsed, leaving a gap of over 43.5 miles (70 km). The impact of these events was so dramatic that, on site of the collapse, a British glaciologist declared he was "stupefied" while an Argentinean colleague broke into tears.

Media from around the world reported these historic events. After all, in a timeframe of only a few weeks, the outline of part of the Antarctic continent had been changed forever. Experts believed that both of these occurrences were largely the result of increased warming of the Antarctic Peninsula's climate. The press viewed these events as proof of the beginnings of a planetwide catastrophe. It was a sign that global warming was melting the Antarctic ice cap at previously unsuspected rates.

THE SLEEPING GIANT AWAKENS

In early 1998, three years later, another ice shelf collapsed. The Wilkins Ice Shelf, with an area of 425 square miles (1,101 sq km), shattered, creating thousands of icebergs.

More recently, in March 2002, a huge floating shelf known as Larsen B collapsed and splintered into icebergs. With a thickness of 722 feet (220 m), the Larsen B shelf had likely been around for 12,000 years. It was larger than the state of Rhode Island. Its weight was around 720 billion tons (653 billion metric tons)—the equivalent of about 12 trillion twenty-two-pound (10 kg) bags of ice cubes! It wasn't until two years later, in September 2004,

These aerial photographs taken between February and March of 2002 show different aspects of the final stages of the collapse of the Antarctic Peninsula's Larsen B shelf. Over a period of thirty-five days, a 2,020-square-mile (3,250 sq km) area of ice disintegrated, setting thousands of icebergs adrift in the sea. The collapse, a result of fifty years of sharp temperature rises on the Antarctic Peninsula, has reignited fears of global warming and its possible catastrophic effects.

that scientists discovered that Larsen B's shattering had had an effect similar to pulling a cork out of a bottle. Once Larsen B was no longer an obstacle, ice floes that fed the shelf began moving toward the sea, slowly melting along the way. Glaciologists were further concerned by the fact that with the passage unblocked, other glaciers were flowing into the ocean more than six

Disappearing Krill

Changes in sea ice cover have many repercussions. The changes are already affecting the population of krill in the Southern Ocean, for example. Krill are a big topic of discussion among Antarctic researchers. This is because these tiny shrimplike creatures are the essential link in the Southern Ocean's ecosystem, providing food for fish, birds, seals, and whales.

Krill measure about 2.5 inches (6.3 cm) each but often travel together in gigantic swarms. One such group tracked by American scientists in 1981 was estimated to weigh 10 million tons (9 million metric tons), which is equivalent to the weight of 143 million humans! Krill is the most abundant animal on the earth. Scientists were very concerned, however, when an important study published in 2004 revealed that quantities of Antarctic krill have diminished as much as 80 percent since the 1970s.

Krill are the foundation of Antarctica's marine food chain and an important topic in current oceanographic research.

The reason for this sharp decline is believed to be the reduction in sea ice due to warmer temperatures, particularly in the Antarctic Peninsula where krill breed. Krill feed on microscopic plankton and algae that live under the surface of sea ice. Operating like a type of nursery, the ice protects algae from harsh surrounding elements, allowing it to thrive. Melting ice has thus resulted in a decrease in algae as well as fewer krill.

continued on following page

> Already the greatly diminished amount of krill has been linked to the decline in some species of Antarctic penguins and the inability of some whales to make a comeback after decades of uncontrolled hunting led to their near extinction. It is estimated that a full-grown blue whale eats about four million krill a day (the equivalent of 4 tons [3.6 metric tons] of shrimp). Its one-day intake would satisfy a single human being for four years.

times faster than before. Aside from acting as a braking system for glaciers, shelves keep warmer sea air at a distance, which leads to less melting. As glaciers flowed more rapidly down to the sea, more water would enter the oceans and lead to an increase in sea levels.

Recently, glaciologists have also observed that because of global warming, 75 percent of the 400 mountain glaciers on the Antarctic Peninsula are diminishing. In the last fifty years alone, over 502 square miles (1,300 sq km) of the peninsula's sea ice has disappeared. They calculated that if all the ice melted, global sea levels would rise by 3 feet (1 m). What worries some scientists even more, though, is the increasing changes global warming is having on West Antarctica.

West Antarctica has long been considered stable because of its icy temperatures. However, recent research shows that as warmer seawater melts the ice from underneath, glaciers on West Antarctica are also beginning to collapse. In fact, global warming is affecting ice closer to the South Pole than ever before recorded. In 2001, the Intergovernmental Panel on Climate Change (IPCC) published a report stating that the collapse of the West Antarctic ice sheet was unlikely to occur during the twenty-first century. Since then, however, three large ice shelves on West Antarctica have been consistently losing ice, which breaks off into the sea. Studies carried out since 2000 have concluded that melting Antarctic ice is currently contributing at least 15 percent to the annual global sea level rise of 0.08 inches (2 mm). Based on these findings, glaciologists now think it is possible that the West Antarctic ice sheet could indeed break up during the next 100 years. Such an occurrence would cause global sea levels to rise as much as 16.5 feet (5 m) and would have severe effects on coastal areas throughout the world.

THE FUTURE

Despite recent events in the Antarctic, scientists continue to have more questions than answers about the mysterious white continent. Research shows increased breaking off of ice shelves and increased melting, as well as rising sea temperatures. As it warms, the seawater expands, taking up more volume and causing oceans to rise. Meanwhile, less ice and more ocean water can further increase the effects of global warming.

> ## See How It Cracks
>
> Ted Scambos, a researcher with the National Snow and Ice Data Center (NSIDC) at the University of Colorado, and a team of collaborators have researched how ice disintegrates. One major theory, called the meltwater theory, proposes that melted ice produced by warmer summer temperatures leaks into tiny cracks in Antarctic ice, causing them to widen. When the meltwater refreezes in the winter, it expands, causing the ice around it to fracture. Meanwhile, warmer ocean currents beneath the surface of ice erode the bottom of ice shelves and seep into cracks from below. Such developments add enormous pressure to ice, leading it to crack and eventually break off or shatter into the sea.

This is because dark surfaces (such as water) absorb more of the sun's light and heat than bright ones (such as ice and snow), which reflect the sun's rays back toward the atmosphere.

The causes of such melting are almost certainly the effects of global warming. Yet scientists aren't sure if these Antarctic changes are the results of human-produced greenhouse gases, local warming trends (such as the increasing temperature of parts of the Southern Ocean, particularly in the area around the Antarctic Peninsula), or a combination of both. What is certain is that the Antarctic Peninsula has warmed approximately 4°F (15.6°C) since the 1950s. Moreover, the summer leading up to the collapse of the Larsen B Ice Shelf was the warmest since Antarctic temperatures were first recorded.

Another unanswered question is whether the Antarctic continent is losing more ice than it is gaining. In recent years, scientists have observed increased snowfall in some parts of East Antarctica. Satellite measurements have shown that East Antarctic central ice thickened at an average rate of 0.7 inches (1.8 cm) per year between 1992 and 2003. Some scientists believe this snow is already a result of global warming, since increased evaporation of moisture leads to increased precipitation. The accumulation of more snow would cause ice sheets to grow thicker and heavier. The increased weight of this ice could actually push the solid rock of the continent's foundation farther down toward the bed of the ocean. Furthermore, more ice, which serves as a cooling influence, could actually offset the effects of warming trends. For instance, sea levels could actually fall in some places if moisture is locked up as ice.

For now, scientists are keeping their eyes on the Ross Ice Shelf, the famous West Antarctica shelf that is the size of Texas. Some worry that it could soon show signs of melting. If this occurred, the entire West Antarctic sheet could begin to slide into the Southern Ocean, where its ice would melt. The result would be a rise of 20 feet (6 m) in global sea levels. However, other

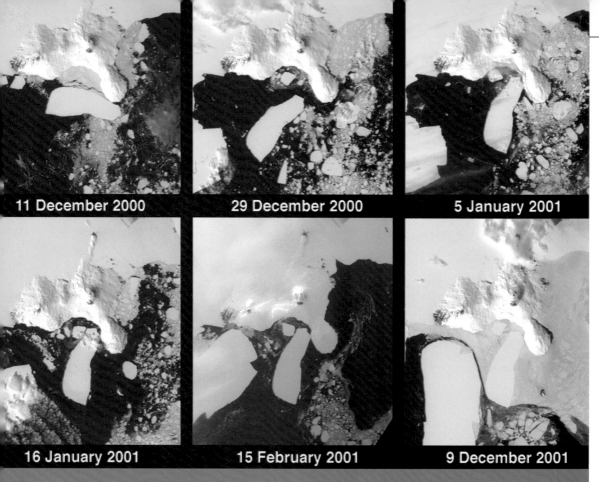

These satellite images show iceberg movements and changes in sea ice between December 2000 and December 2001. They reveal icebergs breaking off from the Ross Ice Shelf, migrating, and forming a barrier that has affected wind patterns and ocean currents. These grounded icebergs also make life difficult for penguins that live near the shelf by increasing the distance between their breeding grounds and the open sea where they catch their food.

scientists claim that it would take fifty years of a warming trend as extreme as the current one in the Antarctic Peninsula for the Ross Ice Shelf to begin to break up.

Ultimately, scientists still have much to discover about how oceans, ice, and the atmosphere interact with one another and the possible consequences for the planet. Scientists working in Antarctica view each core that is pulled from the ice, each shelf that collapses, each iceberg that splinters as various pieces in a gigantic jigsaw puzzle. Each event or discovery offers a small but essential clue to how the white continent works. As the pieces are fitted together and the scientific community comes closer to understanding global warming and the long-range effects it will have on our lives, governments will need to take steps to help minimize damage to the planet.

Glossary

algae Plant-related organisms that live in water.

Archimedes' principle A scientific rule stating that an object partially or totally immersed in liquid is buoyed up by a force equal to the weight of the displaced fluid.

botanist A scientist specializing in the study of plants.

buoyancy The capacity to remain afloat in a liquid or to rise in air or gas.

constellation A cluster of stars grouped together by astronomers for classification purposes.

desolate Deserted, abandoned.

displace To move or shift from the usual place or position.

ecosystem An ecological community together with its environment, functioning as a unit.

gangrene Decay and death of body tissue, often occurring in a limb, caused by insufficient blood supply.

geologist A scientist who studies the origin, history, and structure of the earth.

global warming The gradual heating up of the earth's climate, primarily due to an excess of man-made gases being released into the air.

greenhouse gases Gases, such as carbon dioxide, that upon being released into the air, function like a greenhouse by trapping heat on the earth.

GLOSSARY

krill Tiny shrimplike creatures that live in great numbers in the Southern Ocean and are the favorite food of Antarctic birds, seals, and whales.

meteorologist A specialist who studies processes in the earth's atmosphere that cause weather conditions.

oxygenate To supply with oxygen.

philosopher A scholar who reflects upon and studies various aspects of the world in an attempt to acquire knowledge and wisdom.

plankton Small or microscopic organisms that float or weakly swim in great numbers on the surface of freshwater or salt water.

scurvy A disease caused by deficiency of vitamin C (found in fruits and vegetables), characterized by soft gums, bleeding under the skin, and extreme weakness.

strait A narrow channel joining two larger bodies of water.

windchill index A still-air temperature that would feel the same on human skin as a given combination of wind speed and temperature.

For More Information

Canadian Polar Commission
Suite 1710, Constitution Square
360 Albert Street
Ottawa, ON K1R 7X7
Canada
(888) 765-2701 or (613) 943-8605
Web site: http://www.polarcom.gc.ca/english/antarctic/
    ccar.html

Intergovernmental Panel on Climate Control
IPCC Secretariat
c/o World Meteorological Organization
7 bis Avenue de la Paix
C.P. 2300, CH-1211
Geneva 2
Switzerland
Web site: http://www.ipcc.ch/

National Oceanic and Atmospheric
    Administration (NOAA)
14th Street & Constitution Avenue NW
Room 6217
Washington, DC 20230
(202) 482-6090
Web site: http://www.noaa.gov/

FOR MORE INFORMATION

National Snow and Ice Data Center (NSIDC)
449 UCB, University of Colorado
Boulder, CO 80309-0449
Web site: http://www-nsidc.colorado.edu/iceshelves/

Office of Polar Programs (OPP)
The National Science Foundation
4201 Wilson Boulevard
Arlington, VA 22230
(800) 877-8339 or (703) 292-5111
Web site: http://www.nsf.gov/dir/index.jsp?org=OPP

United States National Ice Center
Federal Building #4
4251 Suitland Road
Washington, DC 20395
(301) 394-3100
Web site: http://www.natice.noaa.gov

WEB SITES

Due to the changing nature of Internet links, the Rosen Publishing Group, Inc., has developed an online list of Web sites related to the subject of this book. This site is updated regularly. Please use this link to access the list:

http://www.rosenlinks.com/eet/anme

For Further Reading

Apte, Sunita. *Polar Regions: Surviving in Antarctica*. New York, NY: Bearport Publishing, 2005.

Armstrong, Jennifer. *Shipwreck at the Bottom of the Sea: The True Story of the* Endurance *Expedition*. New York, NY: Crown Books, 1998.

Bancroft, Ann, and Nancy Loewen. *Four to the Pole! The American Women's Expedition to Antarctica, 1992–1993*. North Haven, CT: Shoe String Press, 2001.

Currie, Stephen. *Antarctica*. San Diego, CA: Lucent Books, 2004.

Dewey, Jennifer Owens. *Antarctic Journal: Four Months at the Bottom of the World*. New York, NY: Harper Collins, 2001.

Matsen, Bradford. *An Extreme Dive Under the Antarctic Ice*. Berkeley Heights, NJ: Enslow Publishers, 2003.

Schlesinger, Arthur Meier, and Fred L. Israel, eds. *Race for the South Pole: The Antarctic Challenge*. New York, NY: Chelsea House, 1999.

Taylor, Barbara. *DK Eyewitness Guides: Arctic and Antarctic*. New York, NY: DK Publishing, 1995.

BIBLIOGRAPHY

Antarctica.org. "Polar Challenges: The World of Antarctic Ice." Retrieved August 2005 (http://www.antarctica.org/Hp_Uk/).

Antarctic Connection. "Glaciology." Retrieved August 2005 (http://www.antarcticconnection.com/antarctic/science/glaciology.shtml).

BBC News. "Antarctic's Ice 'Melting Faster.'" February 2, 2005. Retrieved August 2005 (http://news.bbc.co.uk/1/hi/uk/4228411.stm).

British Antarctic Survey Web site. "About Antarctica." Retrieved August 2005 (http://www.antarctica.ac.uk/).

CIA. "Antarctica." In *The World Fact Book*. 2005. Retrieved August 2005 (http://www.cia.gov/cia/publications/factbook/geos/ay.html).

Cool Antarctica. "Antarctica—Pictures, Information and Travel." Retrieved August 2005 (http://www.coolantarctica.com).

Government of Canada. "Taking Action on Climate Change." Retrieved August 2005 (http://www.climatechange.gc.ca/english/).

Gulf of Maine Research Institute. "Antarctica." Retrieved August 2005 (http://octopus.gma.org/surfing/antarctica/antarctica.html).

National Snow and Ice Data Center (NSIDC). "Antarctic Ice Shelves and Ice Shelves in the News." Retrieved

August 2005 (http://www-nsidc.colorado.edu/iceshelves/).

PBS. "Warnings from the Ice." Retrieved August 2005 (http://www.pbs.org/wgbh/nova/warnings/).

United Nations Environmental Programme. Overview Geo-2000: "The Polar Regions." Retrieved August 2005 (http://www.unep.org/geo2000/ov-e/0010.htm).

USA Today. "Cold Science: Understanding Polar Ice." Retrieved August 2005 (http://www.usatoday.com/weather/resources/coldscience/iceexplore.htm).

INDEX

A

Amundsen, Roald, 17–20
Antarctica, 4
    exploration of, 11–20
    history of, 8–21
    name of, 11
    physical features of, 6–7, 14, 16, 17
    population of, 22
    possession of, 24–26
    research in, 20–21, 22–25, 35–36, 38
    stations in, 20–21, 22–23, 25, 27
    time zone of, 16
    tourism and, 27
Antarctic ice caps, 4
    cracks in, 52
    drilling of, 43
    ice types on, 28–30
    melting of, 5–6, 21, 46–55
    physical features of, 6–7, 26–30
    and rising sea levels, 5–6, 39–41, 50, 51, 54
Antarctic Treaty, 25–26
Archimedes' principle, 39, 40

B

British Antarctic Survey, 4, 46

INDEX

C
Cook, Captain James, 12–13

D
da Gama, Vasco, 11
Davis, John, 14
de Gerlache, Adrien, 15–17

G
glacial ice, 28–29
glaciology, 41–45, 46, 48–50
global warming, 7, 36–41, 50, 52, 55
Gondwana, 8–9
greenhouse gases, 36–37

I
icebergs, 30–32, 39, 41, 55
    proposed uses for, 31
ice cores, 45
ice shelf, 29–30
ice streams, 29
Intergovernmental Panel on Climate
    Change (IPCC), 5, 51

L
Larsen B, 47, 48, 53

M
Magellan, Ferdinand, 11
McMurdo Station, 20

melting ice caps, 5–6, 21, 46–55
    effects of, 49
Mount Erebus, 14

O
ozone layer, depletion of, 24

R
Ross Ice Shelf, 30, 53, 54
Ross, Sir James Clark, 14–15

S
Scott, Robert Falcon, 17–20
sea levels
    effects of, 5, 7, 41, 51
    estimates on, 5
    falling, 53
    rising, 4, 5–7, 39–41, 50, 51, 54
seal and whale hunting, 13–14, 15
Siple Dome, 43
Southern Ocean, 32–34, 49, 52
South Pole, 7, 16, 17, 38, 51
    race for, 18–20
supercontinent, 8–9

V
Vespucci, Amerigo, 11
Vinson Massif, 7

W
Wilkins Ice Shelf, 47

ABOUT THE AUTHOR

Michael A. Sommers is a writer and journalist with university degrees from McGill University, Montréal (BA in literature) and the École des Hautes Études en Sciences Sociales, Paris (masters in history and civilizations). Having grown up in Canada, he has a strong affinity for things glacial and a high tolerance level for sub-zero temperatures. Writing this book significantly increased his awareness about the uniqueness as well as the importance of the Antarctic continent and its impact on the future of the planet.

PHOTO CREDITS

Cover, pp. 4–5 © Jeffrey Kietzmann/National Science Foundation; pp. 1, 22 © Michael Van Woert/NOAA NESDIS/ORA; p. 8 © Chris Danals/National Science Foundation; p. 9 courtesy of the USGS; p. 12 © Library of Congress, Prints and Photographs Division; p. 13 © State Library of Tasmania; p. 16 © Zee Evans/National Science Foundation; p. 19 (top left) © Time Life Pictures/Getty Images, Inc.; pp. 19 (bottom left), 42 © AP/Wide World Photos; p. 19 (top right) © Mr. Steve Nikolas/NOS/NGS; p. 20 © Fox Photos/Getty Images, Inc.; p. 21 © Ann Hawthorne/Corbis; p. 23 © Galen Rowell/Corbis; p. 24 © NASA Goddard Space Flight Center; p. 26 © Library of Congress Geography and Map Division; p. 29 © National Geographic; p. 33 © Library of Congress Geography and Map Division; p. 35 © Frans Lanting/Corbis; p. 37 © Michael Gilbert/Science Photo Library; p. 38 © Jim Reed/Corbis; p. 44 © Alexander Colhoun/National Science Foundation; p. 46 © Melissa Rider/National Science Foundation; p. 48 © Reuters/Corbis; p. 49 © Peter Johnson/Corbis; p. 54 © NASA Jet Propulsion Laboratory.

Designer: Thomas Forget; Editor: Liz Gavril
Photo Researcher: Hillary Arnold

MAR 2 6 2008

SAYVILLE LIBRARY
11 COLLINS AVE.
SAYVILLE, NY 11782